INTERNATIONAL CENTRE FOR MECHANICAL SCIENCES

COURSES AND LECTURES - No. 28

ELWYN R. BERLEKAMP
BELL TELEPHONE LABORATORIES
MURRAY HILL, NEW JERSEY

A SURVEY OF ALGEBRAIC CODING THEORY

LECTURES HELD AT THE DEPARTMENT
FOR AUTOMATION AND INFORMATION
JULY 1970

UDINE 1970

SPRINGER-VERLAG WIEN GMBH

Originally published by Springer-Verlag Wien New York in 1972

ISBN 978-3-211-81088-0 ISBN 978-3-7091-4325-4 (eBook)

DOI 10.1007/978-3-7091-4325-4

Preface

This paper is a survey of coding theory written for mathematicians and statisticians who have some familiarity with modern algebra, finite fields, and possibly some acquaintance with block designs. No knowledge of stochastic processes, information theory, or communication theory is required.

Most of this paper is base on my lectures given to the Department of Automation and Information of the International Center for Mechanical Sciences in Udine during early July, 1970. Some of the lecture material, which dealt with a review of the basic properties of finite fields, is not included in this paper. Instead the paper includes a small amount of additional material dealing with extensions of the results covered in the lectures.

A number of text-books and survey papers dealing with coding theory have already been published, and one might reasonably ask why I have written yet another at this time. While it is true that this survey is a bit more up to date than its predecessors, it is also true that something upwards of 75 per cent of the material in this survey can be found in a more complete, more general, or more rigorous form in my 1968 book, Algebraic Coding Theory.

There are two major reasons for my writing this survey: 1) To reach a new audience, which cannot reasonably be expected to learn about coding theory from the vast, disorganized literature available elsewhere and 2) to help further clarify the subject for myself. Any further clarification generally helps in identifying additional research problems, and the present case is no exception. It was while writing this paper that I first formulated the problem solved in Berlekamp (1971). In this limited sense, the second objective of writing this paper has already been achieved.

It is my pleasant duty to record here my thanks to the Authorities of CISM, especially to Prof. L. Sobrero and G. Longo, for their invitation to present these lectures. I would also like to thank the students at Udine for their tolerance of my unfortunate tendency to lapse into a hurried dialect of English which is only semi-intelligible to Europeans, and to Prof. R.G. Gallagher of MIT, whose concurrent series of lectures at Udine provided relief for both me and the students.

Udine, July 1970

Elwyn R Berlekamp

1. Introduction to Coding Theory.

A binary block code is a set of n-dimensional vectors over the binary field. Each of these vectors is called a codeword. All n-dimensional vectors, including those not in the code, are called words. The number of components in each word, n, is called the block length. The code is said to be linear iff the sum of every pair of codewords is another codeword. Of course, the sum of two binary vectors is computed component by component without carries, according to the rules $0 + 0 = 1 + 1 = 0$, $0 + 1 = 1 + 0 = 1$. The number of codewords in a linear code is a power of 2, say 2^k and the code may be conveniently specified as the row space of a $k \times n$ binary generator matrix, G whose rows may be taken as any k linearly independent codewords. Alternatively, the code may also be defined as the null-space of a $(n-k)\times n$ binary parity-check matrix, H which satisfies

$$HG^t = 0 \quad .$$

The row space of H is the orthogonal complement of the row space of G.

The n dimensional row vector C is a codeword

iff it satisfies the equation

$$Hc^t = 0 \quad .$$

If R is any n-dimensional binary vector, the weight of R, denoted by $w(R)$ is the number of ones among its n coordinates. The notion of weight is an important concept which distinguishes coding theory from classical linear algebra. Indeed, three of the most important questions about the code defined as the null-space of any given parity-check matrix H are the following :

1. What is the weight distribution of the codewords? In other words, for each $i = 0, 1, \ldots, n$, how many codewords have weight i?
2. What is the weight distribution of the cosets? The set of all 2^n binary vectors may be partitioned into 2^{n-k} cosets, where x and y are in the same coset iff $Hx^t = Hy^t$.

   The weight of a coset is defined as the minimum weight of all 2^k vectors in the coset. For each $i = 0, 1, \ldots, n$, how many cosets have weight i? More generally, for each of the 2^{n-k} cosets, what is the distribution of the weights of the 2^k words within that coset?

3. The decoding problem may be defined as follows: Given an arbitrary n- dimensional vector, R, find a minimum weight vector among the 2^k vectors in the coset containing R. In most cases of interest, 2^k is so large that an exhaustive search of the vectors in the coset is not feasible. A more practical decoding algorithm is then desired.

These questions arise naturally when codes are used to obtain reliable communication over less reliable channels. A typical communication system using linear block coding operates as follows : First, a "message source" generates a message consisting of k bits. It is usually assumed that all 2^k possible k - bit messages are equiprobable. The encoder selects one of the 2^k n-bit codewords, C, corresponding to the k-bit message. Each of the n bits of the selected codeword, C, is then transmitted across a noisy channel. The rate of the code, R, is defined by $R = k/n$. In some codes, k of the n transmitted bits may be identified as message bits and the other $n-k$ bits are check bits. In such cases, R is the fraction of the transmitted bits which are message bits.

The channel adds to the transmitted codeword C an error word, E, whose n components are selected independently at random from a distribution which is 0 with probability $q > 1/2$ and 1 with probability $p < 1/2$. It is generally assumed that p is rather small, so that most (but probably not all) of the components of E will be zero. In fact, the probability that the channel adds in a particular error word, E, is given by $p^w q^{n-w}$ where $w = w(E)$. The probability of various error patterns is clearly a monotonic decreasing function of their weights.

The receiver receives the n-bit received word $R = C + E$, which is the sum of the transmitted codeword and the added error word. The decoder then attempts to decompose R into the codeword C and the error word E. Of course, the decoder cannot do this with certainty; any of the 2^k codewords is always possible, no matter which R is received. Corresponding to each possible codeword C is a possible error pattern, $E = R - C$ (+).

Since $HC^t = 0$, it follows that $HE^t = HR^t$. For given R the decoder may compute HR^t. This $n-k$ dimensional vector is called the <u>syndrome</u>. It is often denoted by s^t. The equation $s^t = HE^t$ is satisfied by 2^k possible error patterns. Although each of these error patterns is possible, the lighter ones are more probable. The decoder's best strategy is to choose a

(+) Of course, for binary vectors $R - C = R + C$. Although we are presently concerned almost exclusively with the binary case, we shall use $-$ instead of $+$ on certain occasions to make it easier for the reader who wishes to extend this discussion to nonbinary channels.

minimum weight x from among the 2^k solutions of the equation $Hx^t = s^t$. The selected x is called the leader of the coset with syndrome s. If the decoder decides that x was the error pattern, then the decoded word is $C' = R - x$. If $x = E$, then $C' = C$ and the decoder is said to be correct. If $x \neq E$ then $C' \neq C$ and a <u>decoding error</u> has occurred.

A decoding algorithm which finds a minimum weight x for each of the 2^{n-k} possible syndromes s is called a <u>complete decoding algorithm</u>. Since a complete decoding algorithm will decode correctly iff the channel error pattern is a coset leader, the probability of a decoding error with a complete decoding algorithm depends directly on the weight distribution of the coset leaders.

A decoding algorithm which finds the minimum weight x only for some subset of the 2^{n-k} possible syndromes is called an <u>incomplete decoding algorithm</u>. The simplest incomplete decoding algorithm decodes only the zero syndrome, $s = 0$. The leader of the coset with this syndrome is obviously the all-zero vector. This simple decoding algorithm is correct iff the error word is $E = 0$, which happens with probability q^n. However, there are two distinct ways in which this algorithm can be incorrect. It can commit a decoding error, which will happen iff the error word is a nonzero codeword, or it can commit a decoding failure, which will happen whenever the error word is not a codeword. Decoding failure is generally considered preferable to decoding error. The probabilities of decoding error and decoding failure for the simple algorithm which decodes only the zero syndrome are directly related to the weight distribution of the codewords.

A more sophisticated incomplete decoding algorithm, called <u>bounded distance decoding,</u> decodes iff the received word lies in a coset which has weight less than or equal to some number denoted by t, which is called the algorithm's <u>error-correction capability.</u> If the received word lies in a coset of weight $> t$, then the bounded distance decoding algorithm fails to decode, detecting that at least $(t + 1)$ of the code's bits were changed by the channel noise. The probability of the bounded distance decoding algorithm being correct is equal to the probability that the channel error word has weight no greater than t, which is clearly given by

$$\sum_{i=0}^{t} \binom{n}{i} p^i q^{n-i} .$$

It turns out that the probabilities of error and failure for all bounded distance decoding algorithms may be determined from the weight distribution of the codewords, although the expression and its derivation, originally due to MacWilliams (1963), are too complicated to be given here. They may be found on p. 399 of Berlekamp (1968).

Of course no decoding algorithm can have error-correction capability t unless all words of weight $\leqslant t$ lie in distinct cosets. This will happen iff the minimum nonzero weight of the codewords is $\geqslant 2t+1$. For if two words of weight $\leqslant t$ lie in the same coset, then their difference is a codeword of weight $\leqslant 2t$. Conversely, if the code contains a word C of weight w and $0 < w \leqslant 2t$, then we may find x and y such that $C = x - y$, where x and y each has weight $\leqslant t$. Since $x-y$ is a codeword, x and y are in the same coset. From this we conclude that a code has a t-error-correcting decoding algorithm iff its minimum nonzero weight is at least $2t+1$. Of course, if the block length is large the t-error-correcting decoding algorithm may not be feasible in any practical sense.

2. The Gilbert Bound.

Before we actually construct and analyze several specific codes in detail, we will demonstrate the existence of "good" linear codes by a simple argument first introduced by E.N. Gilbert. Suppose we have a linear code with 2^i codewords of block length n and a minimum nonzero weight d which contains a coset of weight $w \geqslant d$. Then we may form an augmented code, whose 2^{i+1} codewords are the union of the 2^i original codewords and the 2^i words in the coset of the original code. The new code is easily seen to be linear and its minimum nonzero weight is still d. The augmented code has 2^{n-i-1} cosets, and if any of these has minimum weight $w > d$, then we may again augment the augmented code without decreasing d. In fact, we may augment the code again and again, each time doubling the number of its codewords, until we obtain a code whose minimum nonzero weight is greater than the weight of any of its cosets. Since each coset of the final code then has a leader of weight $< d$, it follows that the number of cosets cannot exceed the number of words of weight $< d$ and

$$2^{n-k} \leq \sum_{i=0}^{d-1} \binom{n}{i}.$$

It follows that there exists a k-dimensional linear binary code of block length n and minimum nonzero distance d for any values of n, k and d which satisfy this inequality.

Setting $R = k/n$, $u = d/n$, and letting n, k, d go to infinity while R and u remain fixed ($0 < R < 1, 0 < u < 1/2$,), the Gilbert bound becomes

$$R \geq 1 - H(u)$$

where

$$H(u) = -u \log_2 u - (1-u) \log_2 (1-u) \quad .$$

If $u > 2p$, the central limit theorem assures us that in a sufficiently long block of n digits the weight of the channel error word will almost certainly be less than $un/2$, so that if $t = wn/2$, a t-error-correcting decoding algorithm will decode correctly with very high probability. Therefore, on a channel with crossover probability $p < 1/4$, the Gilbert bound assures us that there exists a sequence of codes, having longer and longer lengths but each having the same rate, with the property that the probability of error using bounded distance decoding algorithms approaches zero as the code length approaches infinity.

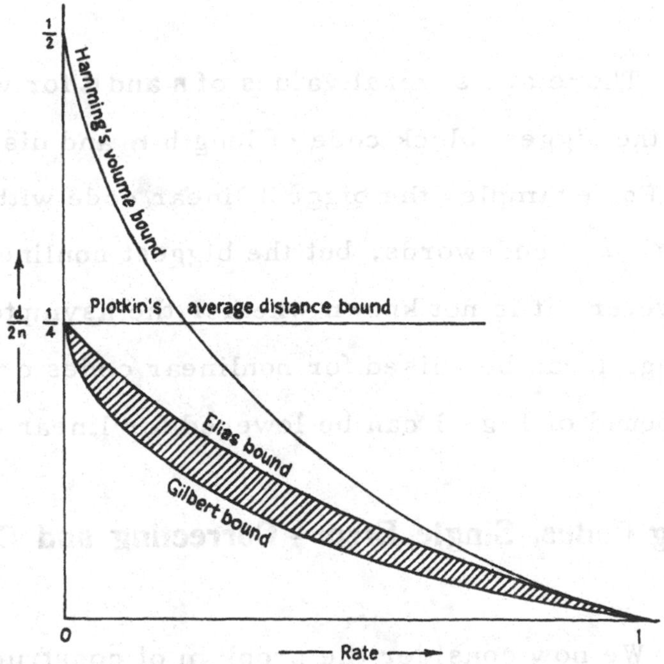

Fig 1

Using more complicated arguments which we shall not reproduce here, it can also be shown that no sufficiently long block code (linear or nonlinear) which has minimum nonzero weight $d = un$ can have more than about $2^{R^*(u)\,n}$ codewords. The functions $R(u)$ and $R^*(u)$ are plotted in Fig. 1. The derivation of $R^*(u)$, originally due to Elias, is given in Chapter 13 of Berlekamp (1968). Tighter upper bounds on the minimum distance of binary block codes of short to moderate lengths have recently been obtained by Johnson (1970), but the new bounds still have the same asymptotic form as the Elias bound.

There are several values of n and d for which it is known that the biggest block code of length n and distance d is nonlinear. For example, the biggest linear code with $n = 15$ and $d = 5$ has 2^7 codewords, but the biggest nonlinear code has 2^8. However, it is not known whether the asymptotic lower bound of Fig. 1 can be raised for nonlinear codes or whether the upper bound of Fig. 1 can be lowered for linear codes.

3. Hamming Codes, Single Error - Correcting and Cyclic.

We now consider the problem of constructing linear codes which are capable of correcting all error words of weight no greater than t. If the code is defined as the null space of the H matrix, and the syndrome is given by $S^t = HR^t$, then the decoder must attempt to find a minimum weight E which solves the equation

$$HE^t = s^t \quad .$$

If we write $E = [E_1, E_2, \ldots\ldots, E_n]$ and $H = [H^{(1)}, H^{(2)}, \ldots\ldots, H^{(n)}]$ where each E_i is a binary number and each $H^{(i)}$ is an $(n-k)$ dimensional column vector, then

$$s^t = \sum_{i=0}^{n-1} E_i H^{(i)} \quad .$$

In other words, the syndrome may be interpreted as the vector sum of those columns of the H matrix corresponding to the positions of the errors.

If all error words of weight zero or one are to have distinct syndromes, it is evidently necessary and sufficient that all sums of 0 or 1 column of the H matrix must be distinct. In other words, all columns of the H matrix must be distinct and nonzero. If the H matrix of a single-error-correcting code has m rows, then it can have at most $2^m - 1$ columns. A linear single error-correcting code of length $n = 2^m - 1$ and dimension $2^m - 1 - m$ is called a Hamming code.

The columns of the parity check matrix of a Hamming code must be the set of all nonzero binary m-tuples. The ordering of these columns is arbitrary, but some orders have certain advantages over others. The most convenient orderings, called cyclic orderings, are based on the cyclic multiplicative group of the Galois field $GF(2^m)$. It is well-known that this field, like all finite fields, contains at least one <u>primitive</u> element, which we call α which satisfies the equation $\alpha^i = 1$ iff i is a multiple of $2^m - 1$. The powers of α include all nonzero field elements, and each may be expressed as a sum of the basis elements $1, \alpha, \alpha^2, \ldots, \alpha^{m-1}$. The representation of each power of α may be uniquely determined from the minimal polynomial of α, which has degree m.

For example, if $m = 4$, then the minimal polynomial of α may be taken as $x^4 + x + 1$. Since α is a root of this polynomial, we have

$$
\begin{aligned}
\alpha^4 &= \phantom{\alpha^3 + \alpha^2 + {}} \alpha + 1 \\
\alpha^5 &= \phantom{\alpha^3 + {}} \alpha^2 + \alpha \\
\alpha^6 &= \alpha^3 + \alpha^2 \\
\alpha^7 &= \alpha^3 \phantom{{} + \alpha^2} + \alpha + 1 \\
\alpha^8 &= \phantom{\alpha^3 + {}} \alpha^2 \phantom{{} + \alpha} + 1 \\
&\;\;\vdots \\
\alpha^{14} &= \alpha^3 \phantom{{} + \alpha^2 + \alpha} + 1 \\
\alpha^{15} &= \phantom{\alpha^3 + \alpha^2 + \alpha + {}} 1 .
\end{aligned}
$$

The ith column of the cyclic Hamming code may be taken

as $H^{(i)} = \alpha^i$, where α^i is represented as an m-dimensional binary column vector, $[A_0, A_1, \ldots, A_{m-1}]^t$, where the A's are the binary numbers determined by the equation $\alpha^i = \sum_{j=0}^{m-1} A_j^{(i)} \alpha^j$. For example, if $m=4$ and $\alpha^4 + \alpha + 1 = 0$, we have

$$H = \begin{bmatrix} 000100110101111 \\ 100110101111000 \\ 010011010111100 \\ 001001101011110 \end{bmatrix}$$

which is often abbreviated as

$$H = [\alpha^1, \alpha^2, \alpha^3, \ldots, \alpha^{15}].$$

The vector C is a codeword iff $HC^t = 0$. If H is the parity-check matrix of a cyclic Hamming code, then this condition is equivalent to

$$\sum_{i=1}^{n} C_i \alpha^i = 0 \quad \text{in} \quad GF(2^m).$$

It is now easy to see that any cyclic shift of any codeword is another codeword. For if $C = [C_1, C_2, \ldots, C_n]$ and $C' = [C'_1, C'_2, \ldots, C'_n]$, where $C'_i = C_{i+j}$, the subscripts computed modulo n, then $\sum_{i=1}^{n} C'_i \alpha^i = (\alpha^{-j}) \sum_{i=1}^{n} C_i \alpha^i$.

Evidently, the new matrix formed by taking any cyclic shift of the columns of H still defines the same code. In fact, the conventional definition of the parity check matrix of a cyclic Hamming code is not the one given above, but rather

$$H = [\alpha^0, \alpha^1, \ldots, \alpha^{n-1}].$$

The coefficients of the words are correspondingly labelled from 0 to $n-1$. The polynomial whose coefficients are the components of a codeword C is denoted by $C(z) = \sum_i C_i z^i$. Clearly $C(z)$ is a codeword iff $C(\alpha) = 0$. But α is a root of a binary polynomial iff that polynomial is a multiple of the minimal polynomial of α, which we denote by $M(z)$. In general, $M(z) = \prod_{i=0}^{m-1} (z - \alpha^{2^i})$. In the present example, $M(z) = z^4 + z + 1$.

Thus, the 2^{11} codewords of the binary cyclic Hamming code of length 15 may be taken as the 2^{11} binary polynomials of degree < 15 which are multiples of $M(z) = z^4 + z + 1$.

4. Linear Cyclic Q - Ary Codes.

In general, the polynomial whose coefficients are a single right shift of the coefficients of $C(z)$ is $z\,C(z)$. If $C(z)$ is a polynomial of degree less than n, interpreted as an n-dimensional vector, then the right cyclic shift of $C(z)$ must have C_{n-1} as the coefficient of z^0 and C_{i-1} as the coefficient of z^i for $i = 1, 2, \ldots, n-1$. The polynomial which represents the right cyclic shift of $C(z)$ is therefore $z\,C(z) \bmod z^n - 1$.

If $C(z)$ is a codeword in a cyclic code of length

then for any i, $z^i C(z) \bmod z^n - 1$ is also a codeword. In a linear code over the finite field $GF(q)$, scalar multiples of codewords are also codewords, so $f_i z^i C(z) \bmod z^n - 1$ is also a codeword. Finally, in a linear code, sums of codewords are also codewords. Therefore, if $C(z)$ is a codeword in a linear cyclic code of length n so is $f(z) C(z) \bmod z^n - 1$, for any polynomial $f(z)$ over $GF(q)$. We therefore reach this conclusion :

> A linear cyclic q-ary code is an ideal in the ring of polynomials over $GF(q)$ modulo $z^n - 1$.

It follows from well-known theorems of classical algebra that this ideal is a principle ideal, and that every codeword is a multiple of some generator polynomial, $g(z)$. This generator polynomial is the monic codeword of least degree, and it is a divisor of $z^n - 1$.

In order to study the various possible linear cyclic codes, we must therefore investigate the divisor of $z^n - 1$ over $GF(q)$. If p is the characteristic of $GF(q)$ and n is a multiple of p, say $n = Np$, then $z^n - 1 = (z^N - 1)^p$. Because $z^n - 1$ has repeated factors in this case, it turns out that most of the cyclic codes are relatively poor. It is therefore usually assumed that n and q are relatively prime. With this assumption, q must be a root of unity in the ring of integers $\bmod n$, and we may denote the multiplicative order of $q \bmod n$ by the integer m. Then n divides $q^m - 1$ but n does not divide $q^i - 1$ for any i smaller than m. It follows that $z^n - 1$ factors completely in $GF(q^m)$ but not in any smaller extension field of $GF(q)$. In $GF(q^m)$, there exists an element α which satisfies the equation $\alpha^i = 1$ iff i is a multiple of n. Such an element is called a primitive n^{th} root of unity. Of course, α is not a primitive field element unless $n = q^m - 1$.

In $GF(q^m)$, $z^n - 1 = \prod_{i=0}^{n-1} (z - \alpha^i)$. Since $g(z)$, the generator polynomial of our linear cyclic q-ary code, is a divisor of $z^n - 1$, it must also factor completely in $GF(q^m)$ For some set of integers $\bmod n$ which we denote by $\bar{K}$,

$$G(z) = \prod_{i \in \bar{K}} (z - \alpha^i).$$

If $\bar{K}$ is taken as an arbitrary set of integers $\bmod n$, then the coefficients of the $g(z)$ defined by this equation will lie in $GF(q^m)$, but they may not lie in the subfield $GF(q)$. The necessary and sufficient condition that all coefficients of $g(z)$ lie in $GF(q)$ is that $\bar{K}$ be closed under multiplication by $q \bmod n$. In other words, $q\bar{K} = \bar{K}$. The quotient of $z^n - 1$ divided by $g(z)$ is conventionally denoted by $h(z)$. It is given by the formula

$$h(z) = \prod_{i \in K} (z - \alpha^i)$$

where K is the set of integers $\bmod n$ which are not in $\bar{K}$. In other words, K and $\bar{K}$ are complementary sets.

The dimension of the linear cyclic q-ary code with generator polynomial $g(z)$ is $n - \deg g$. Since the degree of $g(z)$ is the number of elements in the set $\bar{K}$, the dimension of the code is the number of elements in the set K. This is often denoted by $k = |K|$. A generator matrix for this cyclic code may be taken as

$$G = \begin{bmatrix} g_0 & g_1 & g_2 \cdots & & g_{n-k} & 0 & 0 \ldots 0 \\ 0 & g_0 & g_1 \cdots & & & g_{n-k} & 0 \ldots 0 \\ \cdots & \cdots & \cdots & \cdots & \cdots & \cdots & \cdots \\ 0 \ldots & 0 & g_0 & g_1 \cdots & & & g_{n-k} \end{bmatrix} .$$

A parity check matrix may be taken as

$$H = \begin{bmatrix} h_k & h_{k-1} \cdots & & h_1 & h_0 & 0 & 0 \ldots 0 \\ 0 & h_k & \cdots & h_2 & h_1 & h_0 & 0 \ldots 0 \\ \cdots & \cdots & \cdots & \cdots & \cdots & \cdots & \cdots \\ 0 & 0 \ldots & h_k \cdots & & & & h_0 \end{bmatrix} .$$

The proof that $GH^t = 0$ follows directly from the equation $g(z)h(z) = z^n - 1$. Because of the form of the above parity-check matrix, $h(z)$ is often called the <u>parity-check polynomial</u>. In practice this is sometimes not the most convenient basis of the row space orthogonal to G. In a previous example we considered the binary Hamming code of length 15, for which $g(z)=1+z+z^4$ and $h(z)=z^{11}+z^8+z^7+z^5+z^3+z^2+z+1$. The rows of the parity check matrix given in that example are now seen to be various cyclic shifts of $h(z)$. This is no accident, for all 2^4 words in the null space of G (excepting the all-zero word) are cyclic shifts of $h(z)$. Any set of 4 linearly independent shifts of $h(z)$ could be chosen as the rows of H.

If $g_{aug}(z)$ is a polynomial not divisible by $z-1$, and $g_{exp}(z)=(z-1)\,g_{aug}(z)$, then the pair of cyclic codes with these two generator polynomials are often closely related. The code generated by $g_{exp}(z)$ is called the <u>expurgated code</u> and the code generated by $g_{aug}(z)$ is called the <u>augmented code</u>. The expurgated code is clearly a subcode of the augmented code. Starting from the generator matrix of the augmented code, we may form an <u>extended cyclic code</u> of length $n+1$ by annexing an overall parity check to the cyclic code generated by $g_{aug}(z)$. The generator and check matrices of the extended code are as follows :

$$G_{ext} = \left[\begin{array}{c|c} & \xi \\ G_{aug} & \xi \\ & \vdots \\ & \xi \end{array} \right]$$

where $\xi = g_{aug}(1)$,

$$H_{ext} = \left[\begin{array}{c|c} & 0 \\ & 0 \\ H_{aug} & \vdots \\ & 0 \\ \hline 1 \; 1 \ldots & 1 \end{array}\right] .$$

We may also form the lengthened cyclic code of length $n+1$ from the cyclic code generated by $g_{exp}(z)$. Its generator and check matrices are given by

$$G_{len} = \left[\begin{array}{c|c} & 0 \\ & 0 \\ G_{exp} & \vdots \\ & 0 \\ \hline 1 \; 1 \ldots \; 1 & -n \end{array}\right]$$

$$H_{len} = \left[\begin{array}{c|c} & \gamma/n \\ & \gamma/n \\ H_{exp} & \vdots \\ & \gamma/n \end{array}\right]$$

where $\gamma = h_{exp}(1)$.

It is easily verified that

$$G_{ext} H^t_{ext} = 0 , \qquad G_{len} H^t_{len} = 0 ,$$

and that $G_{len} H^t_{ext} = 0$. It follows that the code obtained by lengthening G_{exp} is identical to the code obtained by extending G_{aug}. In some cases it is possible to show that the extended code is invariant under a transitive group of permutations of its coordinates, and this then enables us to obtain certain information about the weight structure of the original (unextended) code from the following theorem of Prange :

Prange's Theorem :

Let a_j denote the number of codewords of weight j in an original binary code of length n and let A_j denote the number of codewords of weight j in the extended code of length $N = n + 1$. If the extended code is invariant under a transitive group of permutations of its codewords, then

$$a_{2i-1} = \frac{2i\, a_{2i}}{N - 2i} \quad \text{for all} \quad i$$

and the minimum nonzero weight of the codewords must be odd.

Proof :

Consider the $A_{2i} \times N$ matrix whose rows are the codewords of weight $2i$ in the extended code. The leftmost position is taken as the overall parity check, and the a_{2i-1} words which have a one in this position are placed first, as shown in Fig. 2. Since the number of ones in each row of this matrix is $2i$, the total number of ones is $2i\, A_{2i}$. On the other hand, since the code is invariant under a transitive group of permutations of its coordinates, these $2i\, A_{2i}$ ones must be evenly distributed among the N columns. Each column must contain $2iA_{2i}/N$ ones. The first column is known to contain

a_{2i-1} ones, so we have the equation

$$a_{2i-1} = 2i(a_{2i-1} + a_{2i})/N$$

Q. E. D.

Matrix used in proof of Prange's Theorem

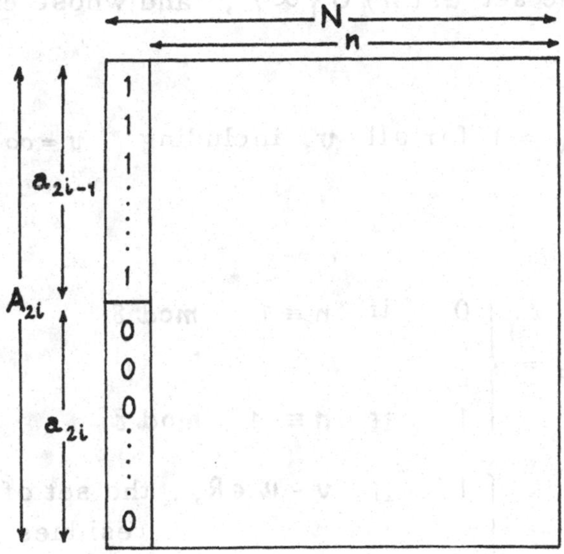

Fig. 2

Sometimes it is convenient to define a linear code directly as the row space of some specified matrix. One interesting example of this approach is the following :

5. Definition of Binary Quadratic Residue Codes.

If $N = n+1$ and n is a prime $\equiv \pm 1 \bmod 8$, then the extended binary quadratic residue code is the row space of the $N \times N$ matrix whose rows and columns are labelled with the N elements in the set $GF(n) \cup \{\infty\}$, and whose entries are given by

$$G_{\infty,v} = 1 \text{ for all } v, \text{ including } v = \infty$$

$$G_{u,\infty} = \begin{cases} 0 & \text{if } n \equiv 1 \mod 8 \\ 1 & \text{if } n \equiv -1 \mod 8 \end{cases}$$

$$G_{u,v} = \begin{cases} 1 & \text{if } v-u \in R_0 \text{, the set of quadratic residues } \bmod n \\ 0 & \text{if } v-u \in R_1 \text{, the set of quadratic nonresidues } \bmod n \\ 0 & \text{if } v-u = 0 . \end{cases}$$

We shall later show that the rows of this $N \times N$ matrix are not linearly independent, and in fact the dimension of the row space of this matrix is $N/2$. The restriction that $n \equiv \pm 1 \bmod 8$ is equivalent to the assumption that 2 is a quadratic residue $\bmod n$. The motivation for labelling the N rows and columns of G with symbols from $GF(n) \cup \{\infty\}$ is the following theorem.

Theorem :

Every extended binary QR code is invariant under the permutation $C_u \leftrightarrow C_{-u^{-1}}$, where the arithmetic operations on the subscripts are performed in $GF(n) \cup \{\infty\}$ with $0 \pm 0^{-1} = \infty$, $\pm \infty^{-1} = 0$.

Proof : We claim that if $u \neq 0$, then

$$G_{-u^{-1},-v^{-1}} = \begin{cases} G_{u,v} + G_{0,v} & \text{if} \quad u \in R_1 \\ G_{u,v} + G_{0,v} + G_{\infty,v} & \text{if} \quad u \in R_0 . \end{cases}$$

This equation follows immediately if $v = 0$ or ∞ or if $u = v$. To check the other cases, we notice that the necessary and sufficient condition that

$$G_{-u^{-1},-v^{-1}} + G_{u,v} = 0 \quad ,$$

is that $(v-u)(-v^{-1}+u^{-1}) \in R_0$. But $(v-u)(-v^{-1}+u^{-1})=\frac{u}{v}\left(\frac{v}{u}-1\right)^2$, so $G_{-u^{-1},-v^{-1}}+G_{u,v}=0$ iff $\frac{u}{v} \in R_0$.

The matrix specified by $G_{-u^{-1},-v^{-1}}$ has the same row space as the matrix specified by $G_{u,-v^{-1}}$, and the matrix specified by

$$\begin{cases} G_{u,v}+G_{0,v} & \text{if} \quad u \in R_1 \\ \\ G_{u,v}+G_{0,v}+G_{\infty,v} & \text{if} \quad u \in R_0 \end{cases}$$

has the same row space as the matrix G.

Q. E. D.

Let us now investigate another property of the matrix defining the extended binary QR codes. If we ignore the row and the column labelled ∞, then each row of the remaining $n \times n$ matrix is a cyclic shift of its predecessor. It follows that this matrix generates a cyclic code, every word of which is a multiple of the polynomial $\sum_{i \in R_0} x^i \bmod x^n - 1$.
The generator polynomial of this linear cyclic code must be the greatest common divisor of the polynomials $f(x)=\sum_{i \in R_0} x^i$ and x^n-1. To find this gcd, we let m denote the multiplicative order of $2 \bmod n$, and work in $GF(2^m)$. The multiplicative group of this finite field is cyclic, of order 2^m-1. Since n divides 2^m-1, $GF(2^m)$ contains a primitive n^{th} root of unity,

called α, such that

$$x^n - 1 = \prod_{j=0}^{n-1} (x - \alpha^j).$$

Since $[f(\alpha^j)]^2 = [f(\alpha^{2j})] = f(\alpha^j)$, it follows that $f(\alpha^j) \in GF(2)$ for all j. If $j \in R_0$, then $f(\alpha^j) = f(\alpha)$, but if $j \in R_1$, then $f(\alpha^j) = 1 + f(\alpha)$ because $f(\alpha) + f(\alpha^j) + 1$ is the sum of all n^{th} roots of unity. By proper choice of the primitive element α, we can assume that $f(\alpha) = 0$, and that the greatest common divisor of $f(x)$ and $x^n - 1$ is given by

$$g(x) = \prod_{i \in R_0} (x - \alpha^i).$$

This equation provides an alternative definition of the augmented QR codes directly in terms of the roots of their generator polynomials. This equation can also be used to define the augmented q-ary quadratic residue code of length n whenever n is prime and q is a quadratic residue mod n.

In general, the equation $g(x) = \prod_{i \in \bar{K}} (x - \alpha^i)$ defines the coefficients of the generator polynomial $g(x)$ in terms of two things: The set $\bar{K}$ and the primitive n^{th} root α. In fact, the important properties of the code are independent of the choice of α. For example, there are two primitive binary polynomials of degree 4 : $x^4 + x + 1$ and $x^4 + x^3 + 1$. Either of these may be taken as the generator polynomial of a cyclic Hamming code of length 15, but the codewords of the second code are easily seen to be obtained by reversing the order of the bits of the corresponding codewords of the first code. In general, if α is one primitive n^{th} root of unity, then any other primitive n^{th} root of unity is of the form $\beta = \alpha^j$, where j is relatively prime to n. The two polynomials $g^{(1)}(x) = \prod_{i \in \bar{K}}(x - \alpha^i)$ and $g^{(2)}(x) = \prod_{i \in \bar{K}}(x - \alpha^{ij})$ generate different cyclic codes, but each codeword of one code may be obtained from a corresponding codeword of the other code by an appropriate permutation of its digits. The permutation is of the form $C_i \longleftrightarrow C_\ell$ where $\ell = ij \bmod n$. A detailed proof is given on page 143 of Berlekamp (1968).

Thus, except for a permutation of its digits, a cyclic code may be defined only in terms of a set $\bar{K}$, without specifying any particular choice of primitive n^{th} root of unity. It is also apparent that if j is relatively prime to n, then the sets $\bar{K}$ and $j\bar{K}$ define equivalent cyclic codes.

In particular, we may define augmented quadratic residue codes simply by stating that $\bar{K} = R_0$, or equivalently, that $\bar{K} = R_1$. In order for this definition to be identical with the previous definition of extended binary quadratic residue codes, which specified each code as a row space of a certain $N \times N$ binary matrix, it may be necessary to choose a certain primitive n^{th} root of unity, α. If a different α is chosen, one gets an equivalent(but not identical)code. Since the augmented and expurgated QR codes are cyclic, if follows that the extended QR code is also invariant under the cyclic shift, $C_u \to C_{u+1}$. Furthermore, the permutation $C_u \to C_{au}$, $a \in R_0$, merely represents a change of the primitive root α. Since the extended QR code is invariant under the permutations $C_u \to C_{-u^{-1}}$, $C_u \to C_{u+1}$, $C_u \to C_{au}$, it must be invariant under the group generated by these permutations, which is the projective unimodular subgroup of the linear fractional group, all of whose permutation are of the form $C_u \to C_v$, where

$$v = \frac{au + b}{cu + d}, \quad \begin{vmatrix} a & b \\ c & d \end{vmatrix} = 1 .$$

Here $a, b, c, d \in GF(n)$, while $u, v \in GF(n) \cup \{\infty\}$. The projective unimodular subgroup of the linear fractional group is transitive (in fact, it is doubly transitive), so we may apply Prange's Theorem to conclude that the minimum nonzero weight of every binary QR code is odd.

In order to compute a bound on the minimum nonzero weight of a quadratic residue code, let $C^{(0)}(x)$ be a codeword of odd weight, let $j \in R_1$, and let $C^{(1)}(x)$ be the polynomial of degree $< n$ defined by the congruence $C^{(1)}(x) \equiv C^{(0)}(x^j) \mod x^n - 1$. $C^{(1)}(x)$ is generally not a codeword, but it is a polynomial which has the same number of nonzero coefficients as $C^{(1)}(x)$. We then observe that since $C^{(0)}(x)$ is a multiple of $\prod_{i \in R_0}(x - \alpha^i)$ and $C^{(1)}(x)$ is a multiple of $\prod_{i \in R_1}(x - \alpha^i)$ the product $C^{(0)}(x)\,C^{(1)}(x)$ is a multiple of $\sum_{i=0}^{n-1} x^i = (x^n - 1)/(x - 1)$. Furthermore, $x - 1$ divides neither $C^{(0)}(x)$ nor $C^{(1)}(x)$, so

$$C^{(0)}(x)\, C^{(1)}(x) \equiv \sum_{i=0}^{n-1} x^i \mod x^n - 1 .$$

But the weight of the product $(x^n - 1)$ cannot excede the product of the weights, so we deduce that the minimum weight, d, of the augmented binary QR code of length n satisfies the inequality $d^2 > n$.

We have thus shown that the extended binary QR codes, each of which has rate $R = 1/2$, have minimum distances approaching ∞ with increasing block length, although the bound is disappointingly weak. The Gilbert bound assures us that there exist arbitrarily long codes of rate $1/2$ whose distances grow linearly with block lengths. However, it is not known whether or not any subsequence of the QR codes has this desirable property.

Although several short QR codes are known to have minimum distances greater than $2+\sqrt{N}$ the actual distances of all sufficiently long QR codes are unknown. The best asymptotic lower bound known for their minimum distances is $\sqrt{N}$. Further known results about QR codes are given in Sec. 15.2 of Berlekamp (1968). (+)

(+) The third printing of this book contains a substantially more up-to-date listing of known minimum distance of QR codes than either of the first or second printings.

Notice how the bound $d > \sqrt{N}$ depends on Prange's Theorem. If $C^{(0)}(x)$ were divisible by $(x-1)$, then we would have $C^{(0)}(x)\, C^{(1)}(x) \equiv 0 \bmod x^n - 1$, from which nothing could be concluded. Thus, without the transitive group of permutations which preserve the extended code, we could only deduce that the minimum odd weight was $> \sqrt{N}$. In some cyclic codes, the minimum odd weight is actually much greater than the minimum nonzero even weight. For example, it is known that the cyclic binary code of length 25 whose generator polynomial's roots include α^i for all i in the set $\bar{K} = \{0, 5, 10, 15, 20\}$ has minimum odd weight 5, but the minimum nonzero even weight is only 2.

6. The Affine Groups.

Although the extended quadratic residue codes are the only codes which are known to be invariant under the projective unimodular subgroup of the linear fractional group, many cyclic codes are known which have extensions that are invariant under a different transitive group of permutations. The length of each q-ary cyclic code whose extension is invariant under this type of group is of the form $n = q^m - 1$. The coordinates of the cyclic code are associated with the nonzero symbols in $GF(q^m)$. The extended code is invariant under the cyclic shift, which fixes 0 and cycles the nonzero coordinates according to the permutation $C_u \rightarrow C_v$, $v = \alpha u$ in $GF(q^m)$. The group of all cyclic shifts is just the multiplicative group of $GF(q^m)$, $C_u \rightarrow C_v$, $v = \beta u$ where $u, v \in GF(q^m)$ and $\beta \in GF(q^m) - \{0\}$. The extended cyclic code may also be invariant under the additive group in $GF(q^m)$, [this group is more conveniently called the translational group], $C_u \rightarrow C_v$, $v = u + \beta$, $v, u, \beta \in GF(q^m)$. If the extended cyclic code is invariant under both cyclic shifts and translations, it is invariant under the group generated by these permutations. This is the little affine group, each of whose permutations is of the form $C_u \rightarrow C_v$, $v = au + b$, $u, v, b \in GF(q^m)$ and $a \in GF(q^m) - \{0\}$. The order of the little affine group is $q^m(q^m - 1)$. In certain cases the code is also invariant under the big affine group, whose general permutation is of the form $C_u \rightarrow C_v$, $v = Au + b$, where now u, v and b are m-dimensional vectors over $GF(q)$, and A is an invertible $m \times m$ q-ary matrix. The order of the big affine group is $q^m \prod_{i=0}^{m-1} (q^m - q^i)$, and it includes the little affine group as a subgroup. The little affine group is doubly transitive for every q, and the big affine group is triply transitive iff $q = 2$.

The following well-known number theoretic result of Lucas (1878) proves useful in characterizing cyclic codes whose extensions are invariant under the little affine group:

Lemma : If p is prime and

$$N = \sum_i N_i p^i \qquad 0 \le N_i < p$$

$$J = \sum_i J_i p^i \qquad 0 \le J_i < p$$

then

$$\binom{N}{J} \equiv \prod_i \binom{N_i}{J_i} \mod p$$

A proof may be found on page 113 of Berlekamp (1968). Since the lemma implies that $\binom{N}{J} \not\equiv 0$ iff $J_i \le N_i$ for all i, it is natural to introduce the notation $J \subseteq N$ iff every digit of the p-ary expansion of J is not greater than the corresponding digit of the p-ary expansion of N. It is easily seen that this relationship is transitive. We say that J is a p-ary descendant of N iff $J \subseteq N$.

Theorem : [Kasami, Peterson, and Lin (1966)]

If q is a power of the prime p, then the extended cyclic code of length $N = q^m$, obtained from the expurgated cyclic code of length $n = q^m - 1$ with generator polynomial given by

$$g(x) = \prod_{i \in \overline{K}} (x - \alpha^i), \quad \alpha \text{ a primitive element in } GF(q^m)$$

is invariant under the little affine group iff the set $\bar{K}$ contains all of its own p-ary descendants.

Proof : Let $C(x) = \sum_{i=0}^{n-1} C_{\alpha^i} x^i$ be an arbitrary q-ary polynomial of degree $< n$ and let $S(C, j)$ denote the value assumed by this polynomial at $x = \alpha^j$. Then

$$S(C, j) = \sum_i C_{\alpha^i} \alpha^{ij} .$$

Clearly $C(x)$ is a codeword of the expurgated cyclic code iff $S(C, j) = 0$ for all $j \in \bar{K}$. If $j \neq 0$, then $0^j = 0$ and $S(C, j)$ is also given by

$$S(C, j) = \sum_{u \in GF(q^m)} C_u u^j .$$

If we define $0^0 = 1$, and

$$S(C, 0) = \sum_{u \in GF(q^m)} C_u u^0$$

then the condition $S(C, 0) = 0$ is necessary and sufficient for C_0 to be the overall parity check.

We now let $T_\Delta(C, j)$ be $S(C', j)$, where C' is the translation of C by an element $\Delta \in GF(q^m)$. Specifically,

$$T_\Delta(C, j) = \sum_{u \in GF(q^m)} C_{u-\Delta} u^j = \sum_{v \in GF(q^m)} C_v (v + \Delta)$$

$$= \sum_{v \in GF(q^m)} C_v \sum_i \binom{j}{i} \Delta^{j-i} v^i .$$

But $GF(q^m)$ has characteristic p, so we may restrict the sum on i to the p-ary descendants of j, interchange the order of summation, and obtain

$$T_\Delta(C,j) = \sum_{i \subseteq j} \binom{j}{i} \Delta^{j-i} S(C,i).$$

Now if $\bar{K}$ contains all of its own descendants, then $S(C,i)$ vanishes for every $i \subseteq j$ whenever $j \in \bar{K}$ and C is a codeword. Similarly, $T_\Delta(C,j) = 0$ for all $j \in \bar{K}$, thus implying that the translation of C is another codeword. Since C might have been any codeword, the extended code is invariant under the translational group.

On the other hand, if $\bar{K}$ does not contain all of its own descendants, then we may select a codeword C for which $S(C,j) = 0$ if $j \in \bar{K}$, $S_j = 1$ if $j \notin \bar{K}$ (+). We may also select a $j \in \bar{K}$ which contains all of its descendants except some $i \subseteq j$ for which $j - i$ is a power of p. For this j, we have

$$T_\Delta(C,j) = \sum_k A_k \Delta^{p^k}$$

where $A_k \in GF(p)$, and $A_k = 0$ if $j - p^k \in \bar{K}$ but $A_k = \binom{j}{p^k}$ if $j - p^k \notin \bar{K}$. If $\ell = \log_p q$, then $N = p^{\ell m}$ and $T_\Delta(C,j) = 0$ for all $\Delta \in GF(p^{\ell m})$ iff the polynomial $x^{\ell m} - x$ divides $T_x(C,j)$. But the degree of $T_x(C,j)$ is less than j, which is not more

---

(+) The proof that in fact there exists such a codeword uses the Mattson-Solomon polynomial, which appears later in this paper, in the proof of the BCH bound.

than $n = q^m - 1$. It follows that there exists a $\Delta \in GF(q^m)$ which translates C into a vector which is not a codeword.

Q. E. D.

This theorem provides a rather easy test to decide whether or not any given extended cyclic code is invariant under the little affine group. We write out the m-digit q-ary expansion of each of the numbers in the set $\bar{K}$, then expand the q-ary expansions into p-ary expansions if necesssary, and then check to see whether $\bar{K}$ contains all of its own descendants. Since multiplication of a number by $q \bmod q^m - 1$ is merely a cyclic shift of the m-digit q-ary expansion of that number, it is natural to represent the elements of $\bar{K}$ in terms of their m- digit q- ary expansions whenever $n = q^m - 1$. The condition that $q\bar{K} = \bar{K}$ merely requires that the cyclic shifts of the m-digit q-ary representation of each number in $\bar{K}$ must be another number in $\bar{K}$.

If j is a number relatively prime to $q^m - 1$, then the permutation $u \to u^j$ in $GF(q^m)$ permutes each extended cyclic code into an equivalent extended cyclic code. This permutation corresponds to multiplying the set $\bar{K}$ by j. Since ancestry is generally not preserved by this permutation, it follows that some extended cyclic codes which are not invariant under the little affine group are equivalent to other cyclic codes which are. Of course, any code equivalent to a code which is invariant under the little affine group is itself invariant under a doubly transitive permutation group isomorphic to the little affine group. In face, the digits of such a code can still be directly associated with the elements of $GF(p^m)$ in such a way as to leave the code invariant under the little affine group. However, this correspondence is not the one assumed in the theorem, which associates the overall parity check with $0 \in GF(q^m)$ and then associates each of the successive powers of the primitive element α. If one associates the successive digits of the code with the successive powers of a wrong primitive element, then the code is not invariant under the translational group because the addition in $GF(q^m)$ is not correctly defined.

In other words, the previous theorem says only that if $\bar{K}$ does not contain all of its own descendants, then the extended code is not invariant under the little affine group if the digits of the code are associated with the elements of $GF(q^m)$ in a particular manner. However, the apparent weakness of this theorem is easily remedied. Since $GF(q^m)$ is unique and its multiplicative group is cyclic, any code invariant under the little affine group with any correspondence between code digits and $GF(q^m)$ is equivalent to a code which is invariant under the correspondence assumed in the theorem. It then follows that if any extended cyclic code invariant under the little affine group has generator polynomial's roots whose $\log_\alpha$ form a set which, when multiplied by some number relatively prime to $q^m - 1$, gives a set which contains all of its own p-ary descendants.

Of all the codes which are invariant under the little affine group, the most highly structured are the generalized Reed-Muller codes.

Definition : The generalized r^{th} order q-ary Reed - Muller code (GRM code) is the extended q-ary cyclic code of length q^m formed from the expurgated cyclic code whose generator polynomial's roots have $\log_\alpha$ in the set

$$\bar{K} = \text{all } a \text{ for which } 0 \leq w_q(a) \leq m - r$$

where $w_q(a)$ denotes the real sum of the digits of the q-ary expansion of the integer a.

The original Reed-Muller codes have $q = 2$, the generalized RM codes allow q to be any prime power. The RM codes include several important subclasses. For example, the RM codes with $r = m - 1$ are the extended Hamming codes.

If j is a p-ary descendant of i and if $q = p^\ell$, then $w_q(j) \leq w_q(i)$. From this it follows that every GRM code is invariant under a little affine group. It is also easily seen that every codeword of the rth q-ary GRM code of length q^m is also in the $(r+1)^{st}$ order q-ary GRM code of length q^m. In other words, the GRM codes of the same length form a nested sequence; each is a subcode of the GRM code of next high-er order.

Let us now examine some of the codewords which are in the rth order RM code but not in the $(r-1)^{st}$. To keep the discussion simple, we first consider only the binary codes, for which $N = 2^m$. Let C be a codeword of the rth RM code, and let $\pi_\Delta(C)$ be the codeword obtained by translating the coordinates of C by $\Delta \in GF(2^m)$. If C is not in the $(r-1)^{st}$ order RM code, then there exists a j with $w_2(j) = r+1$ and $S(C,j) \neq 0$. Similarly, $T_\Delta(C,j) = \sum_{i \subseteq j} \Delta^{j-i} S(C,i)$. But if $i \subsetneq j$, then $w_2(i) < w_2(j)$ and $S(C,i) = 0$. Therefore, $T_\Delta(C,j) = S(C,j)$. It follows that if $C' = C + \pi_\Delta(C)$, then $S(C',j) = S(C,j) + T_\Delta(C,j)$. In other words, the sum of a codeword of the rth RM code and any of its translates is a codeword in the $(r-1)^{st}$ order RM code.

Now suppose that C is the <u>indicator function</u> of a j-dimensional affine subspace of $GF(2^m)$ over $GF(2)$. In other words, C is a vector which has ones in those coordinates corresponding to the points of a certain j-dimensional subspace, and zeroes elsewhere. The weight of C is 2^j. Each of the vectors obtained from C by translating the coordinates of the code by Δ are the affine subspaces which are translates of C . Since the union of a j-dimensional affine subspace of the m-dimensional vector space over $GF(2)$ and any of its translates is a $(j+1)$-dimensional subspace, we conclude that if the rth order RM code contains the indicator functions of all j-dimensional affine subspaces, then the $(r-1)^{st}$ order RM code contains the indicator functions of all $(r-1)$-dimensional affine subspaces. Since the mth order RM code contains all 2^{2^m} binary vectors of length 2^m, it contains the 2^m unit vectors which are the indicator functions of the 0-dimensional affine subspaces. By induction, it follows that the indicator function of every $(m-r)$-dimensional affine subspace of $GF(2^m)$ over $GF(2)$ is a codeword of the rth order RM code. We state the generalization of this result for GRM codes without proof :

<u>Theorem</u> : The indicator function of every $(m-r)$-dimensional affine subspace of $GF(q^m)$ over $GF(q)$ is a codeword of the GRM code of length q^m and order $(q-1)r$.

The 0th order RM code contains only two codewords, the all-zero vector and the all-ones vector, which we denote by $\underset{\sim}{0}$ and $\underset{\sim}{1}$ respectively. The first order RM code contains 2^{m+1} vectors. Except for $\underset{\sim}{0}$ and $\underset{\sim}{1}$ the other $2^{m+1}-2$ codewords are the indicator functions of the $(m-1)$-dimensional affine subspaces.

The first order RM code is the orthogonal complement of the extended Hamming code. The generator matrix of the first order RM code may be written in the following way,

$$G = \begin{bmatrix} 1 & 1 & 1 & 1 & 1 & 1 & \dots & 1 \\ & & & & & & & \\ 0 & 1 & \alpha & \alpha^2 & \alpha^3 & \alpha^4 & \dots & \alpha^{N-2} \end{bmatrix}$$

where each of the last m rows is the m-dimensional binary vector obtained by representing each element of $GF(2^m)$ in terms of some m-dimensional basis over $GF(2)$. We may also write the same G in terms of its rows as

$$G = \begin{bmatrix} \underset{\sim}{1} \\ \underset{\sim}{C}^{(1)} \\ \underset{\sim}{C}^{(2)} \\ \vdots \\ \underset{\sim}{C}^{(m)} \end{bmatrix}$$

Here $1, \underset{\sim}{C}^{(1)}, \underset{\sim}{C}^{(2)}, \dots, \underset{\sim}{C}^{(m)}$ form a basis for the first order code.

The <u>componentwise product</u> of two vectors, $\underset{\sim}{x} = [x_1, x_2, \dots, x_N]$ and $\underset{\sim}{y} = [y_1, y_2, \dots, y_N]$ is defined as $\underset{\sim}{x} \otimes \underset{\sim}{y} = [x_1 y_1, x_2 y_2, \dots, x_N y_N]$. We now assert that for any choice of the binary variables $A_1, A_2, A_3, \dots, A_m$, the component-wise product.

$$\prod_{i=1}^{m} (\underset{\sim}{C}^{(i)} + A_i \underset{\sim}{1})$$

is a unit vector. The proof follows from the fact that the 2^m columns of the last m rows of the G matrix each contains a different m-dimensional binary vector. The column consisting of m ones occurs only once, and each of the other $2^m - 1$ columns contains at least one zero entry. This property is preserved when $\underset{\sim}{1}$ is added to any subset of the rows.

Since the 2^m unit vectors are linearly independent, it follows that the 2^m vectors of the form

$$\prod_{i=1} (\underset{\sim}{C}^{(i)} + A_i \underset{\sim}{1})$$

are linearly independent. Each of these vectors is a polynomial in the m variables $\underset{\sim}{C}^{(1)}, \underset{\sim}{C}^{(2)}, \ldots, \underset{\sim}{C}^{(m)}$. Each of these variables satisfies the binary equation, $\underset{\sim}{C}^{(i)} \otimes \underset{\sim}{C}^{(i)} = \underset{\sim}{C}^{(i)}$. But any polynomial in at most m binary variables may be written as a sum of products, in the form

$$A\underset{\sim}{1} + \sum A_i \underset{\sim}{C}^{(i)} + \sum_{i<j}\sum A_{i,j} \underset{\sim}{C}^{(i)} \otimes C^{(j)} + \sum_{i<j<k}\sum\sum A_{i,j,k} \underset{\sim}{C}^{(i)} \otimes \underset{\sim}{C}^{(j)} \otimes \underset{\sim}{C}^{(k)} + \ldots$$

Since all A's are binary, this form represents an arbitrary polynomial in $\underset{\sim}{C}^{(1)}, \underset{\sim}{C}^{(2)}, \ldots, \underset{\sim}{C}^{(m)}$ as a linear combination of 2^m vectors, each of which is a product of some subset of $\underset{\sim}{C}^{(1)}, \underset{\sim}{C}^{(2)}, \ldots, \underset{\sim}{C}^{(m)}$. (Here $\underset{\sim}{1}$ is the product of the empty subset). It follows that all 2^m componentwise products of the different subsets of $\underset{\sim}{C}^{(1)}, \underset{\sim}{C}^{(2)}, \ldots, \underset{\sim}{C}^{(m)}$ are linearly independent.

The product of any r of the $\underset{\sim}{C}$'s is the indicator function of an $(m-r)$-dimensional affine subspace, and therefore it is in the rth order RM code. The product of any combination of fewer than r of the $\underset{\sim}{C}$'s is the indicator function of a higher dimensional affine subspace, so it is also in the rth order RM code. Thus, we have shown that every polynomial of degree at most r in the $\underset{\sim}{C}$'s is an rth order RM codeword. To show that all r^{th} order RM codewords are of this form, we simply count the number of such polynomials. There are $\binom{m}{i}$ distinct product of i $\underset{\sim}{C}$'s so the number of polynomials of degree at most r in m binary variables is 2^k, where $k = \sum_{i=0}^{r} \binom{m}{i}$.

On the other hand, the dimension of the rth order RM code is $|K| = \sum_{i=0}^{r} \binom{m}{i}$, because there are $\binom{m}{i}$ integers a, $0 \leq a \leq 2^m - 1$, which have $w_2(a) = m - i$. We have thus proved the following result :

Theorem : The codewords of the rth order RM code are the polynomials of degree at most r in m binary variables.

The canonical basis for the rth order RM code consists of the products of all combinations of at most r $\underset{\sim}{C}$'s. In terms of the coordinatization of the code, each of these basis vectors is the indicator function of a certain affine subspace of dimension $\geq m - r$. It is possible to select another basis which consists entirely of indicator functions of $(m-r)$-dimensional affine subspaces. To do this, we simply represent each affine subspace of big dimension as the union of smaller affine subspaces. For example, $\underset{\sim}{C}^{(1)} = \underset{\sim}{C}^{(1)}\underset{\sim}{C}^{(2)} + \underset{\sim}{C}^{(1)}(\underset{\sim}{C}^{(2)}+1)$, so that if $r \geq 2$, the canonical basis vector $\underset{\sim}{C}^{(1)}$, which is the indicator function of an $(m-1)$-dimensional affine subspace, may be replaced by $\underset{\sim}{C}^{(1)}(\underset{\sim}{C}^{(2)}+1)$, which is the indicator function of an $(m-2)$-dimensional affine subspace. The purpose of selecting a basis consisting entirely of indicator functions of $(m-r)$-dimensional affine subspaces is to show that the rth order RM code is the smallest linear code containing all such codewords. It also follows that a coordinate permutation preserves the rth order RM code iff it preserves all $(m-r)$-dimensional subspaces. This group is well-known.

<u>Theorem</u> : The group of all permutations of coordinates which preserve the rth order RM code, $0 < r < m-1$, is the big affine group.

Trivial exceptions occur when $r = 0$ or $m-1$, and in these cases the code is preserved by the full symmetric group of all permutations of coordinates. When $r = 0$, there are only 2 codewords, 0 and 1 when $r = m-1$, every word of even weight is a codeword.

Although we have shown that the rth order RM code contains codewords of weight $d = 2^{m-r}$, we have not yet shown that this is the minimum weight of the nonzero codewords. This fact may be proved in several ways, and the astute reader may be able to devise an elegant proof of his own.

However, we shall adopt a more devious proof, which is <u>constructive</u> from the decoder's point of view. We shall exhibit a decoding procedure and prove that it works whenever the error pattern has weight less than $d/2$.

7. Threshold Decoding.

In general, the decoder wishes to separate the received word, R, into the sum of the transmitted codeword and the channel's error word, E. To this end, he can compute the dot product PR^t for various vectors P. If P is in the null space of the code, then P is said to be a parity check equation and $PR^t = PE^t$. The decoder must attempt to determine E from his knowledge of PE^t for various P in the null space of the code. Since the various components of E are given by UE^t for various unit vectors U, we may rephrase the decoder's problem as the determination of QE^t for certain vectors Q not in the null space of the code from his knowledge of PE^t for P's which are in the null space of the code. The eventual goal is the determination of QE^t for the unit vectors Q, but a good set of subgoals is often the determination of QE^t for certain Q's which have smaller weights than any nonzero vector in the null space of the code.

In the particular case of Reed-Muller codes, the vectors P in the null space of the code are easily characterized. One immediate consequence of the definition of GRM codes is that the orthogonal complement of the rth order GRM code is equivalent to the $((q-1)m-1-r)^{th}$ order GRM code. It follows that the null space of the $((q-1)m-1-r)^{th}$ order contains the indicator function of every affine subspace of dimension However, the null space does not contain the indicator functions of the r-dimensional affine subspaces.

Let S be an r-dimensional affine subspace of $GF(2^m)$ and let $Q(S)$ be its indicator function. The entire space of m-dimensional binary vectors may be partitioned into 2^{m-r} disjoint sets, each of which is a translate of S. We denote the $j = 2^{m-r}-1$ translates which are distinct from S by $T_1, T_2, \ldots, T_j$, and their indicator functions by $Q(T_1)$, $Q(T_2), \ldots, Q(T_j)$. Now since S is an r-dimensional affine subspace and T_i is one of its translates, $S \cup T_i$ is an $(r+1)$-dimensional affine subspace. It follows that $Q(S) + Q(T_i)$ is in the null space of the rth order RM code. Thus, although the decoder cannot determine $Q(S)E^t$ directly, he can determine $Q(S)E^t + Q(T_i)E^t$ for $i = 1, 2, \ldots, 2^{m-r}-1$. For each i, this provides an estimate (+) of $Q(S)E^t$ namely,

$$Q(S)E^t \approx [Q(S) + Q(T_i)]R^t .$$

Clearly this estimate is correct iff $Q(T_i)E^t = 0$.

(+) Since both sides of the subsequent expression are binary, the symbol $\approx$ here means "probably equal" rather than approximately equal".

Following democratic principles, the threshold decoder determines $Q(S)E^t$ by taking a <u>majority vote</u> among these $2^{m-r}-1$ estimates. Specifically, he decides that $Q(S)E^t = 1$ iff $Q(S)E^t + Q(T_i)E^t = 1$ for more than $(2^{m-r}-1)/2$ values of i. Notice that this decision will be incorrect only if more than $(2^{m-r}-1)/2$ affine subspaces T_i each contains an odd number of errors. Since the T_i are disjoint, this requires that the weight of the error pattern be at least 2^{m-r-1}. We state this result as a theorem.

<u>Theorem</u> : If $w(E) < 2^{m-r-1}$ and if PE^t is known for every P which is the indicator function of an $(r+1)$-dimensional affine subspace, and if Q is the indicator function of an r-dimensional affine subspace, then QE^t may be determined by majority vote among the $2^{m-r}-1$ values of PE^t for which $P \otimes Q = Q$.

Once the decoder has computed QE^t for those Q which are indicator functions of r-dimensional subspaces, he can then use the same procedure to determine QE^t for those Q which are indicator functions of $(r-1)$-dimensional subspaces, and continue until he finds UE^t for the unit vectors U. If $w(E) < 2^{m-r-1}$, every election is correctly decided and the decoder's final evaluation of E is correct. It follows that the rth order RM code has minimum distance $d \geq 2^{m-r}-1$. Since all codewords have even weights and the indicator functions of $(m-r)$-dimensional affine subspaces are codewords, it is clear that $d = 2^{m-r}$.

From a practical point of view, threshold decoding is most attractive when r is small. For moderate values of r the number of intermediate subspaces whose parity checks must be computed can be very large.

Notice that threshold decoding succeeds in correcting many error words of weight $> d/2$. If d and N are large, then most error patterns of weight slightly more than $d/2$ will not be evenly distributed among the translates of any r-dimensional subspace S. Less than half of these $2^{m-r} - 1$ translates will contain an odd number of errors and the elections will determine the correct values of $Q(S)E^t$ for all r-dimensional subspaces S.

If m is odd, the $(m-1)/2$ order RM code of length $N = 2^m$ has rate $1/2$, and distance $d = \sqrt{N}$. Thus, the RM codes include a sequence of codes of increasing lengths, each of which has rate $1/2$, and the minimum distances of these codes compare favorably with the best lower bound known for the minimum distances of the extended quadratic residue codes, for which $d > \sqrt{N}$. However, this lower bound on the minimum distance of QR codes appears to be weak, and in all known cases the QR codes are actually as good as or better than the RM codes, when goodness is measured in terms of minimum distance. The threshold decoding algorithm is the major practical attraction of RM codes. No comparably good method for decoding long QR codes is known, even when $w(E) < \sqrt{N}/2$.

The known properties of RM codes enable us to obtain upper and lower bounds for the minimum distances of various other binary cyclic codes of lengths $2^m - 1$. By examining the set of roots of the code's check polynomial, we can find its biggest RM subcode and its least RM supercode. If $w_2(a) \leq m - r \ \forall a \in \bar{K}$ and $m - r' \leq w_2(a) \ \forall a \in K$, then the code is a supercode of the expurgated r^{th} order RM code and a subcode of the r'^{th} order augmented RM code. Its minimum distance, d, is then bounded by $2^{m-r} - 1 \leq d \leq 2^{m-r'}$. In practice, these bounds are often very crude. A stronger technique for bounding the minimum distance of an arbitrary linear cyclic binary code was introduced by Bose-Chaudhuri (1960) and Hocquenghem (1959).

8. BCH Bound.

Theorem : If $\bar{K}$ contains $d_{BCH} - 1$ consecutive elements, modulo n, and if n is relatively prime to q, then there are no nonzero codewords of weight less than d_{BCH} in the linear cyclic q-ary code of length n and generator polynomial

$$g(x) = \prod_{i \in \bar{K}} (x - \alpha^i)$$

where α is a primitive n^{th} root of unity in an extension field of $GF(q)$.

Proof : Mattson-Solomon (1961) : We first obtain an expression for the coefficients of the codeword $C(x)$ in terms of the values $C(\alpha^i)$. It is evident that

$$C(\alpha^i) = \sum_{j=0}^{n-1} C_j \alpha^{ij}$$

Multiplying through by $\alpha^{-i\ell}$ and summing on i gives

$$\sum_{i=0}^{n-1} C(\alpha^i)\alpha^{-i\ell} = \sum_{j=0}^{n-1} C_j \sum_{i=0}^{n-1} \alpha^{(j-\ell)i} = C_\ell \sum_{i=0}^{n-1} \alpha^0$$

because if $j \neq \ell$, then $\sum_{i=0}^{n-1} (\alpha^{(j-\ell)})^i = 0$. Since $\sum_{i=0}^{n-1} 1 = n$, and n is relatively prime to the characteristic of $GF(q)$, we may divide through by n to obtain

$$C_\ell = \frac{1}{n} \sum_{i=0}^{n-1} C(\alpha^i)\alpha^{-i\ell}$$

Now if $C(x)$ is a codeword, then it is a multiple of $g(x)$ and $C(\alpha^i) = 0$ for all $i \in \bar{K}$. We may therefore restrict the sum to the set K to obtain

$$C_\ell = \frac{1}{n} \sum_{i \in K} C(\alpha^i)\alpha^{-i\ell} .$$

There are $d_{BCH} - 1$ consecutive elements not in K. By appropriate choice of b, we may translate this gap to $n - (d_{BCH} - 1)$, $n - (d_{BCH} - 2), \ldots, n - 1$. In other words, we choose b so that for all $i \in K + b$, $0 \leq i \leq n - d_{BCH}$. We then replace i

by $i-b$ to obtain

$$C_\ell = \frac{\alpha^{\ell t}}{n} \sum_{i \in K+b} C(\alpha^{i-b})\alpha^{-i\ell}$$

or

$$C_\ell = \frac{\alpha^{\ell t}}{n} F(\alpha^{-\ell})$$

where $F(x)$ is the Mattson-Solomon polynomial associated with $C(x)$, defined by

$$F(x) = \sum_{i \in K+b} C(\alpha^{i-b})x^i .$$

Since $K+b \subseteq \{0,1,2,\ldots,n-d_{BCH}\}$, the degree of $F(x)$ is at most $n-d_{BCH}$. Therefore, $F(x)$ can have at most $n-d_{BCH}$ zeros, which means that $F(\alpha^{-\ell})$ is nonzero for at least d_{BCH} values of ℓ. The Hamming weight of $C(x)$ is therefore at least d_{BCH}.

Q. E. D.

A change of primitive root multiplies the set K by some number j which is relatively prime to n, and this multiplication will effect the size of the gaps in K. Since the distance structure of the code cannot depend on the choice of primitive root, it follows that the distance of the code is larger than any consecutive gap which occurs in jK for any choice of j relatively prime to n. The least integer which is larger than any gap occurring in jK for any j relatively prime to n is called the Bose distance of the code. The BCH bound implies that the actual distance of the code is at least as large as the Bose distance.

By applying the BCH bound to the GRM codes, we may conclude directly from the definition that the minimum distance of the expurgated $(q-1)r^{th}$ order GRM code of length $q^m - 1$ is at least q^{m-r}. Thus, the binary RM codes satisfy the BCH bound with equality.

For RM codes, the set $\bar{K}$ typically contains many numbers not in the consecutive subset. Some of these numbers have binary expansions which are cyclic shifts of the binary expansions of numbers in the consecutive subset and these numbers are needed to make the code binary. However, other numbers in the $\bar{K}$ for RM codes may have the property that none of their cyclic shifts lie in the consecutive subset. By deleting such elements from $\bar{K}$, we may obtain a higher rate cyclic code which has the same minimum distance.

For example, for the expurgated fourth order RM code of length 63, $\bar{K}$ contains 000000, 000001, 000011, 000101, 001001 and their cyclic shifts, a total of $1+6+6+6+3=22$ numbers. Since 7 is the smallest nonnegative integer having 3 ones in its binary expansion, $\bar{K}$ contains the consecutive sequence $0, 1, 2, \ldots, 6$. By the BCH bound, the expurgated 4th order RM code of length 63 therefore has minimum distance 8. Now notice that one of the cycles in $\bar{K}$, namely 001001 (which represents the triplet 9, 18, 36 mod 63) contains no numbers in the consecutive subsequence. We may thus form a new linear cyclic code of length 63 with $|\bar{K}| = 19$. The new code is called a BCH code. It is a supercode of the 4th RM code. Both codes have distance 8, but the BCH code has 2^{44} codewords while the RM code has only 2^{41}. More generally, BCH codes are defined as follows:

<u>BCH Codes</u> : The augmented q-ary BCH code of length n and designed distance d_d is the linear cyclic code whose generator polynomial's roots have $\log_\alpha$ in the set $\bar{K}$ mod n defined by

$$\bar{K} = \bigcup_i q^i \{1, 2, 3, \ldots, d_d - 1\}.$$

The expurgated q-ary BCH code of length n and designed distance $d_d + 1$ is obtained by annexing $\{0\}$ to the set $\bar{K}$.

Sometimes the Bose distance of a BCH code exceeds it designed distance. For example, the binary BCH code of length 31 and $d_d = 9$ has $\bar{K} =$ 00001 , 00011 , 00101 , 00111 and their cyclic shifts. However, since 01001 is a cyclic shift of 00101 , the Bose distance of this code is eleven.

A q-ary BCH code is said to be <u>primitive</u> iff the block length is of the form $n = q^m - 1$. If $n + 1$ is not a power of q, then the code is said to be <u>imprimitive.</u>

The imprimitive BCH codes include a number of QR codes. For example, with $q = 2$, $n = 17$, $d_{Bose} = 3$, $\bar{K} = \{1, 2, 4, 8, -1, -2, -4, -8\}$ or with $q = 2$, $n = 23$, $d_{Bose} = 5$, $\bar{K} = \{1, 2, 4, 8, 16, 9, 18, 3, 6, 12\}$ In each of these cases, the BCH code is also a QR code, and the actual distance is therefore bounded by $d > \sqrt{n}$, which implies $d \geq 5$. In fact, the actual distance of the QR code of length 17 is 5, but the actual distance of the QR code of length 23 is 7, as we shall show later. At this point, we merely wish to point out that some imprimitive BCH codes have actual minimum distances greater than the Bose distance, and the BCH code with $q = 2$, $n = 17$, $d_{Bose} = 3 < 5 = d$ serves as an example of this point.

On the other hand, most primitive binary codes have actual minimum distances equal to the Bose distance. It is known that $d_{actual} = d_{Bose}$ if $d_{Bose} + 1$ is a power of 2. (In this case the primitive BCH code has a RM subcode which has low weight codewords), or if $d_{Bose} + 1 = 2^{m-1} - 2^{j}$ for some $j > m/2$, or if d_{Bose} divides n, or if $d_{Bose} \leq 11$, or if $n \leq 63$, or if any of certain other conditions holds. It had been conjectured that $d_{Bose} = d_{actual}$ for all primitive binary BCH codes but Kasami-Tokura (1969) then found an infinite number of counterexamples, the first case being $n = 127$, $d_{Bose} = 29$, $d_{actual} = 31$. The Kasami-Tokura counterexamples are based on the following result, which we state without proof :

McEliece's Theorem :

If K does not contain any j-tuple, $x_1, x_2, \ldots, x_j$, for which $\sum_{i=1}^{j} x_i = 0^{(+)}$, then the weight of every codeword in the binary cyclic code generated by $g(x) = \prod_{i \in K}(x - \alpha^i)$ is a multiple of 2^j.

Corollary 1 : All weights in the expurgated binary QR code of length $n \equiv -1 \bmod 8$ or the extended QR code of length $N \equiv 0 \bmod 8$ are divisible by 4.

Proof : In this case, -1 is a nonresidue, the negative or every residue is a nonresidue, and no pair of residues (or non-residues) can sum to 0.

Corollary 2 : All weights in r^{th} order RM code of length 2^m are divisible by $2^{[(m-1)/r]}$.

Proof : Multiply K by -1 so that each integer in K has at most r ones in its binary expansion. In general, $w_2(i + \ell) \leq w_2(i) + w_2(\ell)$, from which it follows that the binary expansion of the sum of any j elements of K has at most jr ones. If $j \leq (m-1)/r$, then $jr < m$, so the sum cannot be

(+) The sum of the x's is taken modulo n.

congruent to 0 mod 2^m-1 because the binary expansion of 2^m-1 has m ones.

For $j=1$ the McEliece theorem is straightforward. In this case, $0 \notin K$, so $0 \in \bar{K}$ and $\alpha^0 = 1$ is a root of every codeword. Thus, $C(1)=0$ in $GF(2)$, which implies that the weight of C is even.

Even for $j=2$, the McEliece theorem is relatively deep. The special case given as Corollary 1 was earlier proved by Gleason, but the general McEliece theorem for $j=2$ was first given by Solomon and McEliece (1966). The generalization to $j \geq 3$ appeared in the thesis of McEliece (1967).

The McEliece theorem enables us to strengthen the bounds on the minimum distances of several codes. According to Corollary 1, all weights of the extended binary QR code of length 24 are divisible by 4 so from Prange's theorem we conclude that the minimum distance of the augmented QR code of length 23 must be congruent to -1 mod 4. Since both the square root bound for QR codes and the BCH bound give $d \geq 5$, it follows that $d \geq 7$. A simple counting argument, $\binom{23}{0}+\binom{23}{1}+\binom{23}{2}+\binom{23}{3} = 2^{11}$ shows that no coset can have weight greater than three, so in fact $d = 7$.

The first Kasami-Tokura code is the primitive BCH code of length 127 and $d_d = 29$, for which

K = 0011101, 0011111, 0101011, 0101111, 0110111, 0111111, 1111111, and their cyclic shifts, making the dimension of the code 43. It is easily verified that any other 7-bit binary number has at least one cyclic shift less than 29. Since every element of K has at least 4 ones in its binary expansion, this BCH code is a subcode of the 3rd order augmented RM code. By Corollary 2 and Prange's theorem, all weights in this code are congruent to 0 or $-1 \bmod 4$. Thus, $d \neq 29$, so $d \geq 31$. But the BCH code is also a supercode of the 4th order RM code, so $d = 31$.

No primitive binary BCH code is known whose minimum distance exceeds the smallest value compatible with both the BCH bound and the McEliece theorem, but a number of other techniques have appeared in the literature for obtaining slight improvements of the lower bounds on the minimum distances of certain imprimitive BCH codes. The work of Mattson-Solomon (1961) and Cerveira (1968) yields conditions sufficient to conclude that $d \geq d_{Bose} + 1$ in certain cases. These conditions, too complicated to be presented here, were later refined by Chien and Lum (1966). Hartmann and Tzeng (1970) obtained separate bounds for the minimum odd weight and the minimum nonzero even weight. A summary and these and other results on the minimum distances of imprimitive BCH codes is given by Hartmann (1970).

The rates and distances of long BCH codes

turn out to be related in a very interesting way. If we consider a sequence of BCH codes of increasing block lengths in which the distances grow linearly with block length (i.e., $d = un$, where u is fixed while d and n go to infinity), then we find that the number of information symbols increases as a power of the block length. More precisely, if we let $I(q, n, d)$ denote the number of information symbols in the q-ary BCH code of length n and designed distance d, then Berlekamp (1968, Chap. 12) has shown that

$$I(2, 2^m - 1, u2^m) \approx n^{s(u)}$$

in the sense that

$$\lim_{m \to \infty} \frac{\log_2 I(2, 2^m - 1, u2^m)}{m}$$

exists and is equal to the singular function $s(u)$ which is plotted in Fig. 3.

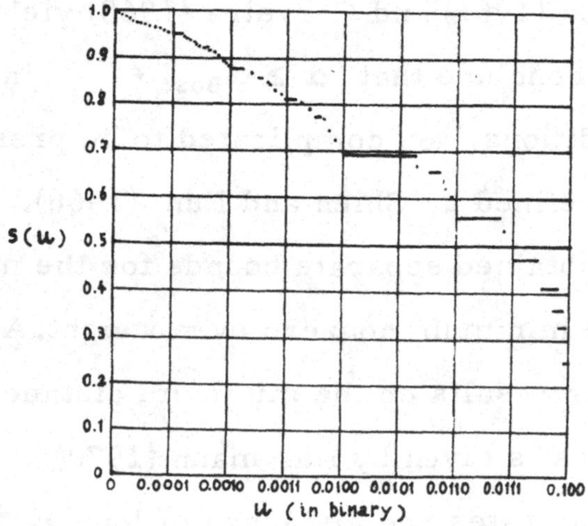

Fig 3

In order to investigate the relationship between the number of information symbols in BCH codes and their actual distances, rather than their designed distances, it is convenient to define the function $\check{I}(q, n, d)$ as the maximum number of information symbols in any of the q-ary BCH codes with actual distance $\geq d$. We then define

$$\check{s}(u) = \limsup_{m \to \infty} \frac{\log_2 \check{I}(2, 2^m - 1, u 2^m)}{m} .$$

It is obvious that $\check{s}(u)$ is a monotonic nonincreasing function of u. It is known that $\check{s}(u) = s(u)$ for an infinite sequence of values of u, some of which are circled in Fig. 3. It is conjectured that $\check{s}(u) = s(u)$ for all u and that the asymptotic ratio between the Bose distance and the actual distance of any sequence of primitive BCH codes approaches one.

Compared with the Gilbert bound, the primitive BCH codes are disappointing. The Gilbert bound asserts that there exist sequences of codes whose distances and information symbols both grow linearly with block length. But if the distance of any sequence of BCH codes grows linearly with block length, then the information symbols grow only as $n^{\check{s}(u)}$, where $\check{s}(u) < 1$.

The same result also holds for sequences of nonprimitive BCH codes of lengths $n = (2^m - 1)/j$, where j remains fixed while m goes to ∞. However, no comparable result is known for sequences of block lengths in which

m = order of $2 \bmod n$ increases much more rapidly than $\log n$.

Instead of investigating the rates of a sequence of BCH codes which have specified distances, we may investigate the distances of BCH codes which have given rates. In particular, if we consider an infinite sequence of extended primitive BCH codes of the same rate, then it is clear from the previous remarks that the ratios of distance/length must approach zero. More precise results have been obtained by Berlekamp (1971), where it is shown that in any sequence of primitive BCH codes in which $\ln R^{-1}/\log n$ approaches 0, the designed distance is given by

$$d = 2n \frac{\ln R^{-1}}{\log n}\left(\frac{\ln R^{-1}}{\log n}\right)^{\left(\frac{\ln R^{-1}}{\log n}\right)}\left(1 + A\frac{\ln R^{-1}}{\log n}\right)$$

where if $\varepsilon > 0$, A may be bounded in the interval $-\frac{1}{2} - \varepsilon < A < \frac{3}{2} + \varepsilon$ for all sufficiently large n. Since δ^{δ} approaches 1 as δ approaches 0, we have the cruder approximation

$$d \approx 2n \frac{\ln R^{-1}}{\log n}$$

This approximation may be obtained from our earlier assertion that $R \approx n^{s(u)-1}$ and the fact that for small u, $s(u) \approx 1 - u/\ln 4$. The behaviour of the singular function $s(u)$ for small u is really quite remarkable. Although $s'(u) = 0$ for almost all positive u, $s'(0) = -1/\ln 4$.

The approximation $d \approx 2n \ln R^{-1}/\log n$ shows that the distances of long BCH codes are much better than RM codes or the best lower bounds known on the distances of QR codes. At $R = 1/2$ the distance of BCH codes grows not as $\sqrt{n}$, but as $n \ln 4/\log n$.

9. Algebraic Decoding of BCH Codes.

From a practical point of view, the major attraction of BCH codes is the existence of an easily implemented decoding algorithm we shall now describe.

The goal of the decoder is to determine the error polynomial, $E(x) = \sum_{i=0}^{n-1} E_i x^i$, where $E_i \in GF(q)$ represents the error which the channel noise adds to the ith digit of the transmitted codeword. Since the received word is the sum of the transmitted word and the channel error word, we have

$$R(x) = C(x) + E(x) .$$

If $i = 1, 2, \ldots, 2t$, then α^i is a root of the generator polynomial of the t-error-correcting BCH code, and since every codeword is a multiple of the generator polynomial, $C(\alpha^i) = 0$ and $R(\alpha^i) = E(\alpha^i)$. Thus, the decoder can determine the quantities $S_i = E(\alpha^i)$ from the equation $S_i = R(\alpha^i)$. From his knowledge of $S_1, S_2, \ldots, S_{2t}$, he then attempts to find $E(x)$.

Since most of the coefficients of $E(x)$ are 0, it is convenient to denote the nonzero coefficients by $Y_1, Y_2, \ldots$. We then have

$$E(x) = \sum_j Y_j x^{e_j}$$

$$S_i = E(\alpha^i) = \sum_j Y_j (\alpha^{e_j})^i .$$

The error word is then characterized by its <u>error locations</u>, $X_j = \alpha^{e_j} \in GF(q^m)$, and its <u>error values</u>, $Y_j \in GF(q)$. In order to determine the error locations, the decoder first attempts to find the coefficients of the <u>error locator polynomial</u>

$$\sigma(z) = \prod_j (1 - X_j z) = \prod_j (1 - z\alpha^{e_j}).$$

The degree of $\sigma(z)$ is the weight of $E(x)$, and the reciprocal roots of $\sigma(z)$ reveal the locations of the errors. In order to find the error values, the decoder first attempts to find the coefficients of the error evaluator polynomial,

$$\omega(z) = \sigma(z) + \sum_j Y_j X_j z \prod_{i \neq j} (1 - X_i z).$$

The degree of ω is not greater than the degree of σ. Once σ, ω and the error locations are known, the error values Y_k may then be determined by evaluating $\omega(z)$ at $z = X_k^{-1}$, which yields the formula

$$Y_k = \frac{\omega(X_k^{-1})}{\prod_{i \neq k}(1 - X_i X_k^{-1})} .$$

Thus, the determination of $E(x)$ from $\sigma(z)$ and $\omega(z)$ is a straightforward procedure; the decoder's major problem is to find the coefficients of $\sigma(z)$ and $\omega(z)$. To obtain relations between $\sigma(z)$ and $\omega(z)$ and the known $S_1, S_2, \ldots,$ we divide ω by σ to obtain

$$\frac{\omega}{\sigma} = 1 + \sum_j \frac{Y_j X_j z}{1 - X_j z} = 1 + \sum_{m=1}^{\infty} S_m z^m .$$

If we define the generating function $S(z) = \sum_{m=1}^{\infty} S_m z^m$ we obtain the equation $(1 + S)\sigma = \omega$. Of course, the decoder does not know all of the coefficients of $1 + S$; he generally knows only $S_1, S_2, \ldots, S_{2t}$. He must therefore attempt to obtain σ and ω from the key equation,

$$(1 + S)\sigma \equiv \omega \mod z^{2t+1} .$$

The best method known for solving this equation is by a sequence of successive approximations, $\sigma^{(0)}, \omega^{(0)}, \sigma^{(1)}, \omega^{(1)}, \ldots, \sigma^{(2t)}, \omega^{(2t)}$, each pair of which solves an equation of the form $(1 + S)\sigma^{(k)} \equiv \omega^{(k)} \mod z^{k+1}$. It is also necessary to give a recursive definition of an integral-valued function $D(k)$ which serves as an upper bound on $\deg \sigma^{(k)}$ and $\deg \omega^{(k)}$. To obtain $\sigma^{(k+1)}$ and $\omega^{(k+1)}$ from $\sigma^{(k)}$

and $\omega^{(k)}$, an additional pair of polynomials, $\tau^{(k)}$ and $\gamma^{(k)}$, are needed. For certain applications of the algorithm to decoding non-BCH codes, it is also convenient (although not strictly necessary) to introduce another function, $B(k)$, which allows the algorithm to break ties accordingly as $B(k)=0$ or 1. The complete algorithm is as follows:

10. An Algorithm for Solving the Key Equation over Any Field.

Initially define $\sigma^{(0)}=1$, $\tau^{(0)}=1$, $\gamma^{(0)}=0$, $D(0)=0$, $B(0)=0$, $\omega^{(0)}(0)=1$. Proceed recursively as follows. If S_{k+1} is unknown, stop; otherwise define $\Delta_1^{(k)}$ as the coefficient of z^{k+1} in the product $(1+S)\sigma^{(k)}$ and let

$$\sigma^{(k+1)} = \sigma^{(k)} - \Delta_1^{(k)} z \tau^{(k)} ,$$

$$\omega^{(k+1)} = \omega^{(k)} - \Delta_1^{(k)} z \gamma^{(k)} .$$

If $\Delta_1^{(k)} = 0$ or if $D(k) > (k+1)/2$, or if $\Delta_1^{(k)} \neq 0$ and $D(k) = (k+1)/2$ and $B(k) = 0$, set

$$D(k+1) = D(k) ,$$

$$B(k+1) = B(k) ,$$

$$\tau^{(k+1)} = z\tau^{(k)} ,$$

$$\gamma^{(k+1)} = z\gamma^{(k)} .$$

But if $\Delta_1^{(k)} \neq 0$ and either $D\ k) < (k+1)/2$ or $D(k) = (k+1)/2$ and $B(k) = 1$, set

$$D(k+1) = k+1-D(k)$$

$$B(k+1) = 1-B(k)$$

$$\tau^{(k+1)} = \frac{\sigma^{(k)}}{\Delta_1^{(k)}}$$

$$\gamma^{(k+1)} = \frac{\omega^{(k)}}{\Delta_1^{(k)}}$$

The most widely known property of this algorithm is that if σ and ω are a pair of relatively prime polynomials which solve the equation $(1+S)\sigma \equiv \omega \bmod x^{2t+1}$, with $\deg \omega \le \deg \sigma \le t$, then $\sigma = \sigma^{(2t)}$ and $\omega = \omega^{(2t)}$. In other words, if the weight of the error word is no greater than t, then this algorithm succeeds in decoding the t-error-correcting BCH code. More general results, which in certain situations allow modifications of the algorithm to correct many error words of weight $t+1$ or $t+2$ in t-error-correcting BCH codes, are given by Berlekamp (1968).

11. Weight Enumerator Theorems.

In the past few years, a large number of results have appeared in the literature concerning the distribution of weights in various codes. Explicit formulas are now known for the number of codewords of each weight in several infinite classes of codes. As mentioned earlier in this paper, these results allow one to compute the probability of decoding failure and the probability of decoding error when one of these codes is decoded by a t-error-correcting decoding algorithm, where t may be any number less than half of the code's minimum distance.

There are so many recent weight enumeration theorems that no attempt will be made to state or prove any of them here. This section is intended only as a guide to the literature. Details of all results which are not otherwise referenced may be found in Chapter 16 of Berlekamp (1968).

Weight enumerators for several classes of codes have been obtained by relatively "direct" methods. Using a variety of combinatorial arguments, several groups of researchers (+) have independently found the formulas for the weight enumerators of a certain class of nonbinary codes which includes all BCH codes for which $n = q - 1$ (++). Applying the classical results of Dickson on quadratic forms over a field of characteristic two, Berlekamp and Sloane (1970) obtained formulas for the weight enumerators of the second-order Reed-Muller codes. Many researchers have determined

(+) Assmus-Mattson-Turyn, Forney, Kasami-Lin-Peterson.

(++) This class of BCH codes was discovered by Reed-Solomon (1960) before the work of Bose-Chaudhuri and Hocquenghem.

several weight enumerators by computer search, the most recent being Chen (1970).

Most of the codes whose weight enumerators have been determined directly, including the 2nd order Reed-Muller codes, have rather low rates. In some sense, the direct methods succeed in obtaining the weight enumerators only because there are relatively few codewords. Fortunately, however, the MacWilliams' identities allow us to obtain the weight enumerator of a high-rate code whenever we know the weight distribution of its orthogonal complement, which is a low-rate code. The MacWilliams theorem is the most fundamental result known about weight distributions. This result states that the weight enumerator of any linear code is determined by the weight enumerator of its orthogonal complement, and it gives an explicit method for calculating either from the other. Another form of the MacWilliams' identities, due to Pless, gives the first r moments of the weight distribution of any linear code in terms of the number of words of each weight $\leq r$ in the orthogonal complement. If one can determine the number of codewords of all but s weights in any given code, and if one can also determine the number of codewords of all weights $< s$ in the orthogonal complement, then it is possible to solve the Pless identities for the complete weight enumerators of both codes.

This indirect approach, which requires an

initial study of the weight distributions of both a given code and its orthogonal complement, has recently proved quite fruitful. Various authors including Pless have used this technique to determine the weight distributions of various quadratic residue codes. Kasami was the first to apply this approach to classes containing an infinite number of codes. The most recent generalizations of Kasami's results are given by Berlekamp (1970). The classes of codes whose weights are enumerated in that paper include all primitive BCH codes of distance ≤ 8 and all primitive BCH codes which contain fewer than $2^{m^2/6}$ codewords of length $2^m - 1$.

Partial weight enumerators are also known for certain other codes. For example, Kasami-Tokura (1970) have obtained formulas for the number of codewords of each weight $< 2d$ in the rth order RM code of minimum distance d.

In spite of all these intriguing results, our knowledge of weight distributions is still far from complete. There are many interesting linear cyclic codes for which not even the minimum distance is known.

References

[1] Berlekamp, E. R. (1968): "Algebraic Coding Theory", MacGraw-Hill, New York;

[2] Berlekamp, E. R., and Sloane, N. J. A. (1970): "The Weight Enumerator of Second Order Reed-Muller Codes" IEEE Trans. Inform. Theory, IT-16 745-751;

[3] Berlekamp, E. R. (1970): "The Weight Enumerators for Certain Subcodes of the Second Order Binary Reed-Muller Codes", Inform. and Control 17 : 485-500;

[4] Berlekamp, E. R. (1971): "Long Primitive Binary BCH Codes Have Distance to appear in IEEE Trans. Inform. Th.;

[5] Bose, R. C., and D. K. Ray-Chaudhuri (1960): "On a Class of Error Correcting Binary Group Codes", Inform. Control, 3: 68-79, 279-290; Math. Rev., 22 : 3619;

[6] Cerveira, A. G. (1968): "On a Class of Wide Sense Binary BCH Codes Whose Minimum Distances Exceed the BCH Bound", IEEE Trans. Inform. Theory, IT-14: 784-785;

[7] Chen, C. L. (1970): "Computer Results on the Minimum Distance of Some Binary Cyclic Codes", IEEE Trans. Inform. Theory, IT-16: 359-360;

[8] Gilbert, E. N. (1952): "A Comparison of Signalling Alphabets", Bell System Tech. J., 31: 504-522;

[9] Hartmann, C. R. P. (1970): "On the Minimum Distance Structure of Cyclic Codes and Decoding Beyond the BCH Bound", Ph D. Thesis, U. of Illinois, 1970; also Co-Ordinated Science Laboratory Report R-458;

[10] Hocquenghem, A. (1959): "Codes correcteurs d'erreur", Chiffres (Paris), 2 : 147-156; Math. Rev., 22: 652;

[11] Johnson, S. (1970): "On Upper Bounds for Unrestricted Error Correcting Codes", to appear;

[12] Kasami, T., Lin, S., and Peterson, W. W. (1966): "Some Results on Weight Distributions of BCH Codes", IEEE Trans. Inform. Theory, IT-12: 274;

[13] Kasami, T., and Tokura, N. (1969): "Some Remarks on BCH Bounds and Minimum Weights of Binary Primitive BCH Codes", IEEE Trans Inform. Theory, IT-15, 408-413;

[14] Kasami. T., and Tokura, N (1970): "On the Weight Structure of Reed-Muller Codes", IEEE Trans. Inform. Theory. IT-16, 752-759;

[15] Lucas, E. (1878): "Sur les congruences des nombres eulériennes et des coefficients différentiels des fonctions trigonométriques,

suivant unmodule premier", Bull. Soc. Math. France, 6: 49-54;

[16] MacWilliams, F. J (1963): "A Theorem on the Distribution of Weights in a Systematic Code", Bell System Tech. J., 42: 79-94; Math. Rev., 26: 7462;

[17] Mattson, H. F. and Solomon, G. (1961): "A New Treatment of Bose-Chaudhuri Codes", J Soc. Indus. Appl. Math., 9: 654-669; Math-Rev., 24B: 1705;

[18] Muller, D. E. (1954): "Application of Boolean Algebra to Switching Circuit Design and to Error Detection", IEEE Trans. Electron. Computers, EC3: 6-12; Math. Rev., 16:99;

[19] Peterson, W. W (1961): "Error-Correcting Codes", The M. I T. Press, Cambridge, Mass.; Math. Rev., 22: 12003;

[20] Reed, I. S. (1954): "A Class of Multiple-Error-Correcting Codes and the Decoding Scheme", IEEE Trans. Inform. Theory, IE-4: 38-49; Math. Rev., 19: 721;

[21] Reed. I. S. and Solomon, G. (1960): "Polynomial Codes Over Certain Finite Fields", J. Soc. Indus. Appl. Math., 8: 300-304; Math. Rev. 23B: 510;

[22] Solomon, G. and McEliece, R. (1966): "Weights of Cyclic Codes", J Combinatorial Theory", 1: 459-475; Math. Rev., 38: 1940;

[23] Tzeng, K. K. and Hartmann, C. R. P. (1970): "On the Minimum Distance of Certain Reversible Cyclic Codes", IEEE Trans. on Inform. Theory, IT-16: 644-645.

Contents

Printed in the United States
By Bookmasters